636.4
ARTLEY
LIVING Living with pigs

Artley, Bob.
Living with pigs

MAR 2 6 2015

Living with
PIGS

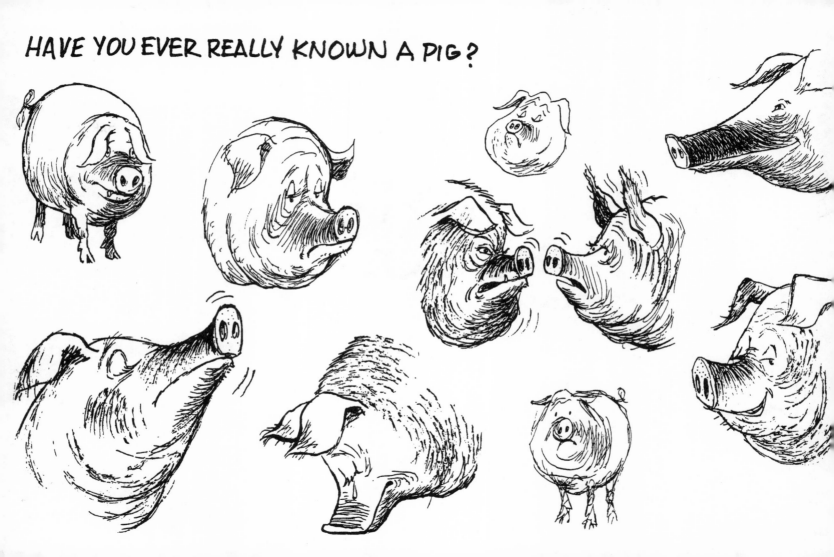

Living with PIGS

Written and illustrated by

Bob Artley

Foreword by Dave Flint

PELICAN PUBLISHING COMPANY
Gretna 2003

For Dean, Reggie, David, Dan, Rob,
and all those family farmers who have known the smell of pigs

Copyright © 2003
By Bob Artley
All rights reserved

The word "Pelican" and the depiction of a pelican are trademarks
of Pelican Publishing Company, Inc., and are registered
in the U.S. Patent and Trademark Office.

Library of Congress Cataloging-in-Publication Data

Artley, Bob.
  Living with pigs / written and illustrated by Bob Artley ; foreword by Dave Flint.
     p. cm.
  ISBN 1-58980-104-0 (pbk. : alk. paper)
  1. Swine. I. Title.
  SF395 .A75 2003
  636.4--dc21

2002152357

Printed in the United States of America

Published by Pelican Publishing Company, Inc.
1000 Burmaster Street, Gretna, Louisiana 70053

CONTENTS

Foreword	7
Preface	9
Acknowledgments	10
Pigs Were My Live Models	13
Farrowing Time	17
Pig Chores	21
Slopping Pigs	31
Corncobs	37
Pigs as Individuals	39
Pigs as Comfort-Seekers	43
Transporting Pigs	45
The Smell of Pigs	47
Side Effects	53
Escape Artists	55
Butchering	57

Pigs in Editorial Cartoons   61
A Family Affair   67
Corn and Hogs   73
Keeping Involved   81
Omnivores   83
Problems of Containment   89
A Sense of Fun   91
Making It All Worthwhile   95

FOREWORD

In the late 1930s and early 1940s, nearly every family farm raised some hogs. They were usually farrowed in the warm spring months, finished on homegrown corn, and sold eight or nine months later. These pigs were raised in dirt lots, where they enjoyed the opportunity to wallow in rainwater or water from a tipped trough. I began a career as a vocational agriculture instructor in 1953 and taught for thirty-five years. During this time I encouraged FFA (Future Farmers of America) members to consider raising pigs, as most farms could provide space for a one- or two-litter project.

Living with pigs provided many young men and women the opportunity to develop the responsibility of caring, keeping financial records, and exercising their decision-making skills and allowed them the chance to compete on judging teams. Many of these students continued their educations and went on to become hog buyers, feed-company reps, country bankers, and lawyers, and many did return to the farms to raise more pigs.

The 1950s saw a concentrated movement toward developing the "meat-type hog." The short, fat, early maturing hogs that were blue-ribbon winners of the 1940s became the white-ribbon winners of the 1950s: they were just too fat for a calorie/cholesterol-watching consumer.

Progress in changing the fat pig to the leaner, well-muscled pig was quite rapid, thanks to the leadership of many Iowans.

Later in the 1980s and 1990s, with larger farm machinery and the capacity it provided a single farm to grow more acres of corn and soybeans, the emphasis moved away from pasture-raised hogs to confinement-raised hogs. Scientists proposed that, with proper rations, ventilation, farrowing crates, and nurseries, pigs could be produced in large numbers with less labor. Thus, there was a movement to large confinement buildings containing hundreds of pigs.

Now, a potbellied semi can haul more pigs in one load than were raised in an entire year on the family farms of the '40s and '50s. As you drive the country roads in Iowa today, you see vast fields of corn/soybeans growing where farmsteads once stood—the buildings are gone, the pigs and other livestock are gone, and the people are gone. These buried farmsteads are memory graveyards for a time that is past and shall never return. This new generation of consumers are interested in "pork, the other white meat," but they don't want the aroma of pig farms infiltrating their cars as they travel—yet they also like Canadian bacon and pepperoni on their pizzas and the pork burger on the grill, not to mention Iowa chops!

Living with pigs in the past was a time-consuming, dirty, smelly job, but for those of us who remember those days, we do so with mixed fondness.

DAVID FLINT

Retired Vocational Agriculture Instructor and FFA Advisor

PREFACE

Pigs have never been my favorite farm animal. But a good portion of my life has been involved with those fascinating, maddening creatures. This was especially true of my young years on the farm.

Pigs were necessary in our farm economy, as they were a fairly fast-maturing animal (from farrowing to market). A sow could produce eight to ten or more piglets each season, and we marketed the corn we grew through them.

I don't pretend to know all there is to know about the hog industry. Maybe that is why I'm so negative toward these factory-farm hog confinements that seem to be taking over today.

But my life with pigs, as limited as it was, allowed me to view them as fellow creatures on this earth, and I want to see them treated well in their short life with its tragic end. Even as a kid on the farm I enjoyed eating pork, but I avoided being present when the unlucky pig was put to death so that we might feast.

ACKNOWLEDGMENTS

The drawings in this book have all appeared before in various forms. I want to thank those publications that first published them: *Worthington Daily Globe* and *Agri News*. I also thank *Extra Newspaper Features* and *Mugwump Marketing* for their syndication services. Many of the drawings subsequently appeared in some of my books published by Iowa State University Press, now being republished by Voyageur Press.

I also want to thank my brother, Dean; my son Rob; my cousins Bob and John for their interest and input; and Dave Flint for his foreword.

Last but not least, I am indebted to my wife, Margaret, for her encouragement and help; Jim Davies, who translated the handwritten script into typed text onto disk; and the staff at Pelican for their patience, understanding, and help.

Living with PIGS

PIGS WERE MY LIVE MODELS

Most serious art students are provided with live, professional, nude models in order to learn to draw the human form. Later, at the University of Iowa, I had a class in which I was provided this opportunity. It was invaluable training.

However, when first learning to draw from life, pigs were my models.

Ever since I started wielding a pencil, I drew every chance I got. But I drew what I saw in my imagination. However, when Virginia Allinson, a close family friend, took some of my sorry samples to show to Ding Darling, he set me straight.

Ding, whose editorial cartoons appeared daily on the front page of the *Des Moines Register*, was my idol from afar. His drawings were, to me, what cartoons should look like, and I would try copying them, as well as making some of my own. I decided at an early age that I wanted to be a cartoonist like Ding.

When our friend brought back the report from her meeting with Ding, I found that his stern advice was that I should draw from life if I was going to learn to draw. And he sent back some pencils and a sketchbook and suggested I use the farm animals as models.

The big problem being, except when they were asleep, they never held their position. One had to sketch fast, which of course was good training.

This I set about doing studiously, sketching our horses, cows, chickens, and pigs every chance I got. Since the pigs in their pens were usually more available, they were the ones that I drew the most.

The fact was, except when I caught them asleep, they were constantly on the move, providing valuable lessons in sketching rapidly. At such times I would have to concentrate on an ear, a snout, or a leg and then quickly move on to some other part of my model's anatomy.

This early training came in handy when, in the university live-model classes, we were required to make quick-action sketches as the models changed their positions every few seconds.

Thus, my early drawing sessions with pigs on our farm were invaluable lessons in my learning to draw. I feel I owe those pigs, as well as my mentor, Ding, who set me on the right path.

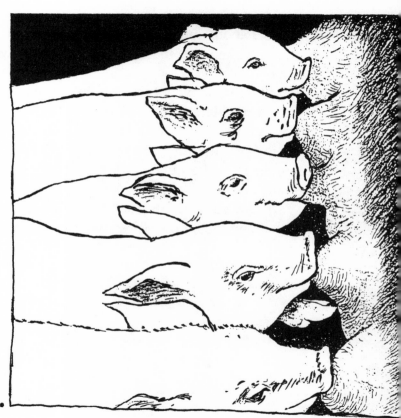

EACH PEN HELD A SOW AND HER LITTER OF EIGHT OR TEN PIGLETS. AS EACH PIG WAS BORN IT SOON FOUND ITS WAY TO ITS FIRST MEAL AT THE MOTHER'S SIDE.

FARROWING TIME

Spring on the farm was an exciting time. Everything green was springing to life, and new life was popping up everywhere.

The swallows, killdeers, robins, house wrens, meadowlarks, field sparrows, orioles, turtledoves, blackbirds, and finches were all returning from their winter places in the sun. They were establishing territories, building nests, and starting their families.

In the barnyard new life was also burgeoning: new calves, baby chicks, ducklings, goslings, and baby pigs.

My first observance of the birth of baby pigs opened my eyes to the mysteries of this miracle, the first chapter to my sex education on the farm.

When the hog house was cleaned, disinfected, and made ready with individual farrowing pens bedded with clean bright oat straw, it was actually an attractive, even cozy, place.

Since this was before electricity came to our farm, we had no heat lamps to help keep the piglets warm during a cold snap. Newly farrowed piglets that were at risk were gathered together into a basket, covered with a gunnysack, and put behind the kitchen range until they were warmed up and could be returned to their mother's side. I can remember how we would confuse the small grunting sound of the warmed-up pigs with the sound of the coffee percolating on the stovetop.

There were times when a cold baby pig was near death and Dad, holding it in his hands in front of the open oven door, massaged its bluish little body back to a healthy pink.

THERE WAS SUCH A SHORT TIME WE COULD PLAY WITH THE CUTE LITTLE PIGLETS —

ALL TOO SOON THEY OUTGREW THEIR CUTE STAGE AND BECAME...

...HOGS!

PIG CHORES

    Our chores on the farm seemed to be excessively devoted to tending to the pigs. They had to be fed twice a day (before we had automatic hog feeders). And these feeding times involved a lot of physical labor, which we boys took for granted when we would try to be men and carry our share of the load.
    The hog-feeding floor was not near the corncrib, so the ear corn had to be carried across the farmyard about a hundred feet or so to be scattered onto the feeding floor, where the pigs would rush in and attack the ears with gusto.
    When we were too small to carry the feed basket like Dad and the hired man did, we hauled the basket of corn on our little wagon and threw the ears over the fence one ear at a time. Then we returned to the corncrib, repeating the task until the required number of basketfuls was delivered.
    In winter, we did not use our little wagon but slid the steel basketful of corn over the snow, going back and forth until we had a smooth path, which we pretended was a train track. This made the chore fun.

As we grew older, we began trying to carry the basket of corn on our shoulders as the men did. This took some struggling and much practice before we were accomplished at it.

Scattering the ears of corn onto the feeding floor from the basket on our shoulders was a tricky business. We had to not only be sure of our footing among the corncobs left from previous feedings, but also be sure of the hungry hogs swarming onto the floor, not in the least heedful of the one feeding them. The process took some fancy footwork if one didn't want to find oneself on the ground with the pigs, corn ears, and cobs. It was best to wait until we were quite mature to try feeding them this way.

When the pigs were sleeping inside the hog house, we tried to be very quiet when approaching the feeding floor. Just one wakeful pig could detect the slightest noise and sound the alarm, bringing the whole population swarming onto the floor and causing us to step lively in scattering the corn.

The same experience was apt to take place when feeding shelled corn or a protein supplement to feeder pigs or sows. Pigs, it seemed, were always in a hurry to eat.

In the fall, during corn-picking season, a load of freshly picked corn was often parked alongside the hog-feeding floor, where the corn could be scooped off to them.

Also in the fall, when the maturing ears hung heavy on the stalk, we would cut a load of corn-laden stalks with a corn knife. This load was fed to the pigs, who not only devoured the ears but also chewed on the green part of the stalks.

AS WE GOT OLDER...

Sometimes, when the corn was mature on the stalks, a portion of a cornfield was closed-off with a temporary, hog-tight fence. The hogs were allowed to go into this fenced-in cornfield and feed to their hearts' content night and day, anytime they wished. This arrangement, with drinking water and a self-feeder with protein supplement nearby, was a particularly efficient way to fatten pigs for market. And it certainly cut down on the chores.

Besides the feeding chores, there were those chores of cleaning the pigpens and putting in fresh straw bedding, which we hauled from stack or mow.

When my son Rob was grown and teaching high school, he supplemented his teacher's salary for a few years by raising market pigs with his uncle Dean. So he too had some of the experiences that his father had years ago, living with pigs.

Years later, when Dean was farming and raising pigs for market by himself, he ground and mixed corn and soybeans and put the mixture into self-feeders. There was a man in our community who owned a portable roaster and would go from farm to farm roasting soybeans. When Dean was grinding and mixing his own hog feed, he included these roasted beans. They smelled like roasted peanuts and made a feed mix the pigs thought to be delicious.

These self-feeders made it possible for the pigs to eat anytime they wished. One of the night sounds during those years was that of the metal lids on the feeders banging as the pigs lifted and dropped them while feeding.

But long before we had self-feeders, there was the exhaustingly tedious job of feeding the

pigs by hand. As described earlier, carrying the ear corn by the basketful and scattering it on the feeding floor, while at the same time dealing with a herd of pigs with voracious appetites, was no small task for man or boy.

Troughs were required to feed the pigs slop or even their dry protein supplement. We built these troughs out of 2 x 6, 2 x 8, or 2 x 10 pine boards. We had to prepare the troughs on occasions because of the pigs used them roughly. More work!

Pigs, by nature, are aggressive, rambunctious, destructive animals. Besides their feeding troughs, their gates, doors, fences, and the buildings they were housed in were in constant need of repair. Then, of course, there was the attention required for each individual pig, such as that which had to be given to inoculate each pig to prevent hog cholera. Inoculating thirty or forty partly grown porkers was no easy job. For this job an appointment had to be set up with our veterinarian, Dr. Schultz of Latimer. He was a very careful, methodical man, setting up his portable equipment with deliberate movements that spoke of efficiency.

I remember how fascinated I was by how he pulled the rubber stoppers out of the serum bottles with his teeth and emptied them into the mouth of the rubber hot-water bottle that hung from a strap around his neck. A rubber tube led from the water bottle to a calibrated syringe. It was from this syringe that he gave a measured portion of serum to each pig.

In the meantime, Dad, Dean, and I would have separated a few pigs from the pen (where the herd had been previously corralled) into a small area where they could be more easily caught.

Squealing and struggling, they were put onto their backs on the *V*-shaped table, where we held them while Dr. Schultz gave each its shot. Then the inoculated pig was lifted, still squealing, over a partition to join the others that seemed to be commiserating with one another over their ordeal.

Another big job was that of castrating the pigs. I hated this much more than inoculating the pigs against cholera. Castration was violence—next to murder in my mind. The poor pig squealed, not only from fear but also from the pain. (They weren't given anesthetics.)

For this, we didn't call the vet. Dad was very proficient at the job, while Dean and I, holding the screaming, struggling pig, suffered in our empathy for the unlucky one. Usually castration was done when the pigs were quite small, making it easier on the pig and certainly on those wrestling with the tussling, squealing victim. After each operation, Dad splashed antiseptic into the wound and released the patient into a pen of straw, where the rest of the former young boars were quietly trying to adjust to their fate. It wasn't easy being a male pig.

In later years, when Dean was raising pigs from farrow to market, he and his wife, Regina, experienced the same ordeals. In addition to the inoculations and castrations, which they had the vet perform, there was the job of clipping the eyeteeth (tusks) and tails from the piglets when they were just days old. This procedure kept the little fellows from biting their siblings' tails off. Regina became quite adept at this clipping process.

We used a fifty-gallon wooden-stave barrel in which to mix the slop. Sometimes it sat next to the fence at the feeding floor with the hog troughs just over the fence, within easy reach. In the winter months, when the troughs were also inside, it was in the alleyway inside the hog house.

There were times when the barrel was wheeled about on a cart designed to carry it. This way, the barrel could be near the milk house where the skim milk, after separating, could be dumped into it. Then the barrel was moved to the granary (where the ground oats and oil meal were added) and then pushed to the stock tank (where water was dipped into the barrel and added to the mixture). Then the swill barrel, in its cart, was wheeled to the feeding floor.

CORNCOBS

Gathering clean, dry corncobs from the hog yard, after the hogs had eaten off all the kernels, was one of our farm chores. These cobs were used as fuel in the kitchen range and were sometimes mixed with coal or wood. Often they were used by themselves as a quick fire, especially in the summertime, to cook a meal and then left to die down with no bed of coals to hold unwanted heat.

The cob pail was a fixture by the stove in the kitchen, along with the coal pail and the stove-wood box.

We started fires in the stove using a corncob set in a can of kerosene. A kerosene-soaked corncob stuck on the end of a stiff wire came in handy, used as a torch when we had to thaw out a frozen water pipe.

When the commercial corn sheller came to the farm, the corn crib was emptied of its ear corn, which was then run through the sheller, separating the kernels from the cobs. The shelled corn was stored or sold and the cobs put into a bin where they could be kept dry and ready for use as fuel. The cob shed was an icon of farmsteads in the early 1900s.

When we had a pet pig and he grew to market size, one of his pleasures was to lie down and let us scratch his belly with a corncob.

PIGS AS INDIVIDUALS

If we got acquainted with an individual pig, we found that there were as many individuals among pigs as there are among humans.

These individuals had their own personalities.

Cousin Bob had one pig that was inquisitive and, if it had been a human, might have become a scientist. Bob had put up a fence to make a space for a temporary hog pasture. The fence was a single electric wire fixed at a height that the pigs could not go under or over without touching. The rest of the pigs went about grazing in their new pasture. But the curious one of the herd came over to explore the boundaries. Seeing the single wire as an unfamiliar kind of fence, he poked his nose against it to investigate. Of course, he got zapped, as was intended. He jumped back, looked at the wire for a while, and again touched his nose to it. Again he got a jolt, and again he jumped back, staring at that mysterious wire. After a third encounter, he apparently decided that the newfangled fence was something he did not wish to investigate further and turned to join the rest at their grazing.

Cousin John's old boar, prone to escaping his pen, was attacked by two farm dogs trying

to corral him. The boar, with his formidable tusks, was not easy for them to handle, until they discovered that he was particularly vulnerable at a place on his rear end.

As the dogs zeroed in on their newfound advantage, the old boar was at their mercy, until he suddenly sat down so the dogs could not get at his vulnerability, showing the logical thinking of this intelligent animal.

Dad gave mom a couple of newborn pigs that were not likely to survive in their overcrowded litter. She named them Spot and Grunt and gave them intensive care as newborns. These two pigs flourished and bonded to her, becoming her pets. When she was out and about the farmyard, they trotted along behind her like a couple of dogs. When they felt they were not getting our mother's attention, they would stand outside the house yard gate and press their noses down on the sidewalk and breath heavily until she would come out to feed them or simply talk to them. I do not remember what happened to those two individuals, but I suppose they went to market with their hog-house peers.

Another piglet-turned-pet was that one my brother and I named Bicycle. We gave this little gilt that name because Dad told us if we raised her to market size we could have the money when she was sold. We determined that with that money we would buy a bicycle, something we had been dreaming of for a long time.

We became attached to our little pet as she grew and we began to dread the time when she would go to market. But our vision of a shiny, new bicycle was so intense that our avarice overcame our feelings, and she went to market.

However, this was probably during an all-time hog-market low. The three cents a pound she sold for was nowhere near enough to buy a bicycle—even during the Depression. So we had no bicycle and no pet.

PIGS AS COMFORT-SEEKERS

Pigs on our farm seemed to be experts at finding a comfortable spot to sleep or relax. In the winter they would snuggle between one another in deep straw for warmth. In summer they sought out shady spots or mudholes they had made.

At one time we found several happy pigs that had jumped into a stock tank to cool off. This practice turned to tragedy one time. When several pigs crowded into a water tank, some went in headfirst and, because they were crowded in, could not right themselves; and when they were discovered, they had drowned.

TRANSPORTING PIGS

We didn't have a pickup truck when I was a kid. When pigs were ready for market, Dad would contact a hog buyer and a stock truck would come to load the twenty or so hogs and take them to the packing plant at Mason City, Iowa.

But when it came to transporting one critter, sow or boar, since we had no pickup truck, the critter was herded into a crate that was then lashed to the side of the car, and we all went on our way together.

This arrangement had negative aspects—smell for one thing. And the prospect of going through town with such a passenger riding on the side of our car did not appeal to Mom in the least.

In earlier days, farm wagons were used to transport pigs to and from market. I remember Grandpa telling of herding a drove of pigs to market down a country road. Knowing the nature of those adventuresome critters, I can imagine what a task it was keeping them on the road and headed in the desired direction. I'm sure it was a task for more than one drover.

THE SMELL OF PIGS

It has been said that the smell of pigs is the smell of money.

I never quite saw it that way. Pig manure is, to me, one of the most foul odors—much worse than chicken, cow, or horse manure. So cleaning a hogpen was an extremely odious job. It left your whole being saturated with that vile smell. Your clothes, your hair, your skin, even your air passages were impregnated with the smell of pigs, often for a long time.

In fact, the smell is so potent that simply walking near a manure-laden pigpen was apt to leave its smell on your clothes.

It is only fair, after having castigated the pigs for its obnoxious smell, to say that there are those who claim that a pig is by nature a clean animal. That, if given a chance, they keep themselves clean. It is true that when we penned up a sow about to farrow, with clean straw in one end of her pen, she would seem to be careful not to foul the nest but would go to her toilet in a corner away from the nest. In fact the hog house at farrowing time was not offensive to the nose.

But as pigs grew, and their numbers caused them to become crowded in their pens, they

seemed to be more slovenly in their habits and not the least bit concerned about their unhygienic lifestyle.

One of the benefits of raising pigs on pasture, with individual houses for shelter, was not having the concentration of hog manure in enclosures in the barnyard, thus eliminating the source of the offending odor—an odor that seemed impossible to get rid of, even with much scrubbing with plenty of soap and water and deodorants.

In animal husbandry, it used to be (and still may be) possible to tell what domestic animals one tended by the aroma emanating from their person, especially in their work clothes. Those who worked closely with horses carried a unique horsy smell; those handling cattle smelled like cattle; those who raised sheep for a living were apt to carry the smell of lanolin about them.

None of these odors are especially offensive to me, unless I'm in a closed space—a theatre, church, or other "polite" gathering—with the person. However, the pig smell, I think, is offensive wherever it is encountered. This, of course, is due to the unsanitary way in which pigs live. I would guess that we humans might smell badly to pigs (but I doubt it) if we lived carelessly with our own excretions.

There are those who say that pigs can be—or they naturally are—clean animals. Having lived a goodly portion of life with those intelligent, unpredictable, often-vicious, adventuresome, maddening, dirty animals, I am not in agreement. A sow in a clean, straw-lined

farrowing pen can be quite clean in her habits—within that pen. But that same creature might go out and lie in a sloppy mess of hog manure, apparently in an effort to cool off on a hot day.

My wife tells of how her mother, a widow on their small farm, with three children, one year raised one pig for their own consumption. The pig's pen and the pig itself were given a daily scrubbing with soap and water. Maybe—just maybe—that pig didn't smell.

SIDE EFFECTS

In the heat of summer, one of the negative aspects of raising any livestock was the flies that generated and became pests all over the farm. And it seemed that the hogpens were especially good generators of the invasive hoard of insects.

The screen door to the house would be covered by flies trying to get in to sample the food being prepared for dinner. Sticky fly tapes and flypaper were dispersed at the back entry of the kitchen. Several fly swatters were used up each season, eliminating those flies that did not end up on the flypaper or sticky tapes hanging from the ceiling.

My wife speaks of how, when growing up on the farm, she would sometimes get those sticky tapes entangled in her hair and what trouble her mother had getting her unstuck from the tape.

A summer mantra we heard constantly was "Shut that screen door!"

ESCAPE ARTISTS

We always thought that, if a pig was not getting out of its pen, it was thinking about it.

There is a story about a salesman who was visiting a farmer in his farmyard when a sow sauntered by. The farmer raised his foot and gave the sow a swift kick.

"Why did you do that?" the salesman asked.

The farmer replied, "She was either coming from or going to some devilment."

All too often that devilment was getting through somewhere in the fence. Dad used to say that if a pig found a hole big enough only for its nose, the rest of it would soon follow.

It seemed most of our life with pigs was trying to keep them in their place.

One August, Dad, Dean, and I planned to go to the Iowa State Fair with Cousin Harold and Uncle George. The morning we were to leave at an early hour, the sows got out and into a cornfield. We spent a lot of precious time in the cornfield, wet with dew, trying to round up those pesky critters, in addition to doing the milking and other morning chores.

We finally, much later than we had planned, were on our way to Des Moines. I do not think we visited the hog barn at the fair that year.

BUTCHERING

Butchering is a grisly business, and when I was a kid, I wanted no part of it. My brother and I made ourselves scarce, appearing at the butchering site only after the hapless critter was a lifeless carcass hanging from the old crab-apple tree behind the corncrib.

I must confess that, while I am fond of smoked ham, bacon, sausage, and an occasional juicy porkchop, if procuring the meat depended on me, I think I might be a vegetarian.

In the early years of our married life, my wife and I did raise chickens for meat one year. I steeled myself and beheaded and helped butcher twenty or more chickens, which we then prepared for storage in the deep-freeze locker in town. For the next several months, our family of six ate chicken—pan-fried, oven-fried, baked, stewed, and fricasseed chicken, chicken and dumplings, and chicken-noodle soup.

In other words, I took part in procuring nourishment for my family as Dad had for his family. In this respect, I felt proud to be a part of the chain of those, who, down through the ages, provided meat for the family table. But I still did not like the grisly business of killing and preparing the meat.

Dad was a gentle man, very concerned, not only for the comfort and well-being of his

family, but also for the well-being of the farm animals for which he was responsible. He butchered hogs, chickens, and beef only because he had to, to provide for his family.

Thankfully, when I farmed for four years after the World War II, I was spared the butchering of our livestock. The meat lockers from Latimer came to the farm, loaded the pork or beef candidates into their truck, and hauled them away. The next time we saw our pig or cow, it was in neat, labeled packages in our rented deep-freeze locker, ready for cooking.

PIGS IN EDITORIAL CARTOONS

In my twenty-plus years of doing editorial cartoons, I found politicians and pigs good subjects—often in the same light.

Pigs are fun to draw, especially when depicted in human context. Actually, a pig is in many ways akin to us humans (sorry, pigs). This is especially true physiologically. If I am not mistaken, pig parts have been considered as a source of replacements for human parts. In fact, hasn't the heart valve of a pig already been used to repair a human heart?

The anthropomorphic characteristics of the pig are quite striking; thus, they become apt candidates as cartoon characters, and I have used them extensively.

In editorial cartooning, we often borrow themes from the classics or from folk tales that are common knowledge in our culture.

In the cartoon on page sixty, I drew upon the theme of George Orwell's *Animal Farm*. In this delightful book, which was a spoof of dictatorship, the farm animals, led by an old boar, rebel against the abusive farmer, drive him from the farm, and assume ownership of the place. The revolution eventually deteriorates and the animal inhabitants discover that their fate, under their esteemed leader, is not better, and is even worse, than before.

At any rate, it was from this literary classic that I was inspired to use the pig—one of my favorite cartoon characters.

If I were drawing editorial cartoons today, with the news of corporate fraud, where greed in high places results in the theft of life savings from workers and trusting investors, I would have plenty of employment for hogs in my cartoons, depicting that despicable side of human nature.

However, pigs are their true selves as nature has made them. They are not scheming, devious, dishonest creatures, as humans too often are.

In my own experience with pigs, I find myself ambivalent toward those fascinating creatures. So I apologize to our fellow creatures, who are so much fun to draw and, in cartoons, can be made to depict our least-admirable traits.

In the words of that wise old pig in the cartoon on page sixty, "I say, just stay away from those filthy people."

MORE ABOUT PIGS
IN THE FOLLOWING PAGES
IN WORDS and PICTURES

A FAMILY AFFAIR

You might say that Mom, Dad, and we boys were the pigs' extended family. From farrowing time in the spring until they went to market, we were all involved, in one way or another, in the pig culture on our farm.

Quite often there was a cold spell at farrowing time. Since we had no electric heat lamps to keep the piglets warm in their nests of straw, those at risk were put into a straw-lined basket, covered with a gunnysack, and put behind the kitchen stove until they had warmed up.

Actually, Mom didn't mind having the cute little fellows in the kitchen. They were in the basket and she enjoyed hearing their tiny grunts and murmuring. Sometimes a particularly stressed little fellow had to be given intensive care—like a massage and a little warm milk from a bottle—until it was strong enough to join its siblings at its mother's side. Mom found this enjoyable. But when the piglets had grown to be big obnoxious hogs, she would just as soon have them in the care of Dad and me and my brothers, in the hog house where they belonged.

When the piglets had not yet grown to adulthood but were still cute and lively, she was again involved. She found herself chasing them out of her flower beds and vegetable garden. They

could do lots of damage to her planted garden with their tough little grisly snouts, even though they were far from being full grown.

In the meantime, Dad and us boys kept busy trying to keep the fences patched to contain the delinquents and see to their inoculations, castrations, and ringing (putting rings in their noses to keep them from rutting up their pasture). There were pens to clean, straw bedding (which we had to haul from the straw stack or from the barn mow) to be placed for their comfort, and trips to be taken by horse and wagon—back and forth to the feed mill in Latimer for their ground feed.

Every so often, as they grew, we had to sort them into different pens according to size and/or sex, keeping aside those who were gilt candidates for breeding next year's crop.

Finally at market time, there was the separating of those that would be sold and those that had more time to shape up before they, too, would be loaded onto the stock truck for their final trip.

Not long afterwards, it was time to clean and disinfect the hog house and prepare it with the individual pens for the next farrowing season.

A SYMBIOTIC RELATIONSHIP

CORN AND HOGS

When I tell people I'm from Iowa and they say, "Oh yes, corn and hogs," it has always been irksome to me. I see my native state as beautiful rolling land between two rivers: the Mississippi and the Missouri. There are the limestone bluffs, caves, and ancient Indian mounds on the east and the prehistoric, wind-created Loess Hills along the western border. It also contains many natural and man-made lakes and parks, including many old-growth forests.

Before the devastating Armistice Day blizzard of 1940, with its suddenly plummeting temperatures that wiped out orchards throughout the state, Iowa was known for its apples.

Iowa has been home to several people of science, literature, and art. Among some of the movies set in the state are *Bridges of Madison County, State Fair, Field of Dreams,* and *Twister.* Meredith Wilson's *The Music Man* was set in the late nineteenth century in his hometown of Mason City, Iowa.

Iowa has produced political figures such as the thirty-first president of the United States, Herbert Hoover, and Henry Wallace, vice president during Franklin Roosevelt's third term. Iowa also was the home of renowned artist Grant Wood, world-famous editorial cartoonist

J. N. "Ding" Darling, as well as Maynard Reece, today's nationally famous wildlife artist. Actor John Wayne is a native Iowan, along with actresses Cloris Leachman, Harriet Nelson, and Donna Reed. Entertainer Johnny Carson and television commentator Harry Reasoner were born in Iowa, as were bass-baritone Simon Estes, singer Andy Williams, and bandleader Glenn Miller. Plus, there is our famous native Iowa author MacKinlay Kantor and advice columnists Ann Landers and Abigail Van Buren. Other native Iowans include plant pathologist and geneticist Norman Borlaug, Olympic gold medalist and college wrestling champion Dan Gable, and polltaker George Gallup. There are many other Iowa men and women of arts and letters, living and dead, who contributed to our society.

But, yes, Iowa does produce corn and hogs, and I was very much a part of that culture.

WE THOUGHT IT GREAT
WHEN THE HOGS
COULD FEED THEMSELVES

KEEPING INVOLVED

My brothers and I were very involved in the pig culture on our farm, whether we chose to be or not. We were born into it as medieval children were born into royalty or peasantry. Dad, in directing our services to these masters, was a kind and compassionate administrator—even though he was a thorough, insistent one.

In retrospect, however, I am glad for the experience that helped teach me the value of hard work and responsibility, but I would not like to repeat it.

Besides the feeding, watering, cleaning, bedding, sorting, and executing the medical procedures, we were kept busy building and repairing pasture and barnyard fences to keep the pigs reasonably controlled. All domestic animals, apparently, yearn to be free. But while three or four barbed (now electric) wires were sufficient to keep a horse or a cow within bounds, just a small hole judged as insignificant would become, to a pig, a challenge to exploit. Fence building with pigs in mind required exactness and attention to the smallest detail.

With intelligence, persistence, and plenty of push, they would soon find a way to muscle through to freedom. Then there was the task of rounding up the escapees and repairing the fence. It was a constant job.

This creamery at Latimer, Iowa, in the early part of this century, was later replaced by a brick building.

A VARIED DIET

OMNIVORES

Pigs, like humans, are omnivores. But, of course, the main diet of the hogs we raised for market in the 1920s and 1930s was corn and oats grown on the farm. During the summer months they were also pastured on clover or alfalfa or rape (a cabbage-related plant).

But in order for them to have a balanced diet, their menu needed to include a protein. Nowadays, soybean meal is the protein of choice. But back when we were raising hogs, the protein available was "tankage," a dry-meal product. It was produced at a rendering plant where the carcasses of dead animals, collected from area farms, were cooked and processed. In addition to the pigs' diet of corn, oats, protein, and green pasture (in season), they were also given a mineral supplement, often purchased from farm-to-farm mineral salesmen. Sometimes a homemade mineral was provided in the form of ashes from the stoves mixed with salt.

Vegetable peelings and other kitchen scraps were also tossed to the pigs. Some cultures use their pigs for garbage consumption, which may account for the biblical admonition against eating pork—considered unclean.

SOMETIMES WE FELT THAT EACH PIG SHOULD BE PUT IN A CRATE—THEN WE'D KNOW THEIR WHEREABOUTS

PROBLEMS OF CONTAINMENT

Just when we thought we had the pigs under control, the fences and pens secure, gates and doors closed, suddenly one or several of those maddening creatures would appear, seemingly out of nowhere, to disrupt any activity we might be engaged in or any we might have planned.

It seemed while we were figuring out how to best contain them, those intelligent creatures were figuring out ways to counter our efforts. It became a battle of wits and the pigs appeared to be winning (as the drawings on pages 84, 85, 86, and 87 imply).

I couldn't help but wonder just how we humans would fare if the roles were reversed. Knowing the intelligence of our adversaries (which, in this case, would be pigs), we probably wouldn't fare too well. If they were designing and building fences or walls to contain us, I have the uncomfortable feeling that we might not be able to escape.

Here again, I am reminded of George Orwell's *Animal Farm,* in which the rebellious animals, led by the intelligent old boar, were able to build a windmill to grind their feed and find other ways to improve the infrastructure of the neglected farm.

PIGS KNEW HOW TO HAVE FUN

A SENSE OF FUN

If pigs could speak our language, I think they might make great entertainers. In fact, when we were raising pigs, it seemed at times that they even possessed a sense of humor.

At any rate, in animal terms, they did seem to enjoy cutting up and playing. It seems that all mammals, including pigs and humans, have been given the blessed capacity to enjoy moments of frivolity. This shared trait is probably why we enjoy our fellow creatures so much.

One year at the Franklin County fair, one of the stage attractions in front of the grandstand featured a troupe of half-grown pigs dressed as clowns. They performed on teetertotters, climbed ladders, and went down a slide, squealing with apparent delight. These crowd pleasers were highly trained young pigs, rewarded at the conclusion of each act by a small morsel the trainer took from his pocket. At some human gatherings, a pig would be coated with grease and then some young athletic-type person chased the frightened pig around the arena, trying to grab onto it and capture the poor creature.

I've also heard of—but never seen—pig races. These events can be quite exciting. I am told bets are even placed on the pigs. It has been observed that the pigs themselves find these races to be fun. Maybe this is so, since pigs do seem to enjoy competing with one another.

MAKING IT ALL WORTHWHILE

MAKING IT ALL WORTHWHILE

Finally, when the chores were all done and we came in from the cold to the warmth and fragrance of the kitchen at suppertime, with the steaming pork roast Mom had just taken from the oven, we did not have to be told twice to "wash up and get to the table."

The brown gravy made from the drippings of the roast, poured over the boiled potatoes, went well with the fresh-baked buns spread with home-churned butter and homemade strawberry jam. It was a feast fit for a king.

But we certainly were not royalty. In fact, we would have been considered the peasant class. However, we felt very rich and privileged to be able to live "so high on the hog." We had no problem entering into the spirit of Dad's table prayer of thanks.

Now, looking back from a perspective of over eighty years, I can appreciate the fact that we were indeed blessed, even though those were the years of the Great Depression. The prices for farm products were depressingly low and cash was scarce, even though our farm home was without modern conveniences (even for that time), we were indeed privileged. We were living on a family farm where we could experience the rewards of our own labor, as long as we didn't